Software Metrics

A How To Guide for Project Staff

David Tuffley

To my beloved Nation of Four
Concordia Domi – Foris Pax

When you can measure what you are speaking about, and can express it in numbers, you know something about it; but when you cannot measure it, when you cannot express it in numbers, your knowledge is of a meagre and unsatisfactory kind: It may be the beginning of knowledge, but you have scarcely in your thoughts advanced to the stage of science.
-- Lord Kelvin

Acknowledgements

I am indebted to the Institute of Electrical and Electronics Engineers on whose work I base this book, specifically IEEE Std 1061.

I also acknowledge the *Turrbal* and *Jagera* indigenous peoples, on whose ancestral land I write this book.

Contents

A. Introduction

This standard defines the standard for software quality metrics methodology. It is meant for people involved with the purchase, development, use, assistance, maintenance or review of software. The standard is especially directed at those measuring or reviewing the quality of software.

This standard is based closely upon IEEE 1061 Standard for a Software Metrics Methodology.

A.1. Scope

This standard supplies a methodology for founding quality requirements and recognising, implementing, analysing and validating process and product software quality metrics. This methodology applies to all software at all stages of any software life cycle structure. Sections 1 through 4 provide, scope, definitions, and background information which is the basis of this standard; all parts of Section 5 are compulsory. Appendices A through D are incorporated for illustrative and reference reasons only.

This standard does not assign specific metrics. However, the appendices include models of metrics together with a finalised example of the use of this standard.

A.2. Objectives

The objectives of this standards is to provide definitive software metrics reference to the following categories of person:

- A purchasing/project manager to identify, state and prioritise the quality requirements for a system.

- A system developer to identify definite features that should be assembled into the software in order to meet the quality requirements.

- A quality audit/assurance/control company and a system developer to review whether the quality requirements are being met

- A system maintainer to aid in change management during product development

- A user to help in distinguishing the quality requirements for a system

A.3. References

[1] IEEE 1061 Standard for a Software Metrics Methodology.

A.4. Definitions & acronyms

Critical range - metric values used to classify software into categories of acceptable, marginal and unacceptable.

Critical value - critical value of a validated metric which is used to identify software which has unacceptable quality.

Direct metric - a metric applied during development or during operations that represents a software quality factor (e.g, mean time to software failure for the factor reliability).

Factor sample - a set of factor values which is drawn from the metrics data base and used in metrics validation.

Factor value - a value (see metric value) of the direct metric that represents a factor.

Measure - to ascertain or appraise by comparing to a standard; to apply a metric.

Measurement - (a) the act of process of measuring (b) a figure, extent, or amount obtained by measuring.

Metrics framework - a tool used for organising, selecting, communicating and evaluating the required quality attributes for a software system; a hierarchical breakdown of factors, sub-factors and metrics for a software system.

Metrics sample - a set of metric values which is drawn from the metrics data base and used in metrics validation.

Software Metrics

Metric validation - the act or process of ensuring that a metric correctly predicts or assesses a quality factor.

Metric value - an element from the range of a metric; a metric output.

Predictive assessment - the process of using a predictive metric(s) to predict the value of another metric.

Predictive metric - a metric applied during development and used to predict the values of a software quality factor.

Process step - any task performed in the development, implementation or maintenance of software (eg., identify the software components of a system as part of the design).

Process metric - metric used to measure characteristics of the methods, techniques, and tools employed in developing, implementing and maintaining the software system.

Product metric - metric used to measure the characteristics of the documentation and code.

Quality attribute - a characteristic of software, a generic term applying to factors, sub-factors, or metric values.

Quality factor - management oriented attribute of software that contributes to its quality.

Quality requirement - a requirement that a software attribute be present in software to satisfy a contract, standard, specification, or other formally imposed document.

Quality sub-factor - a decomposition of a quality factor or quality sub-factor to its technical components.

Sample software - software selected from a current or completed project from which data can be obtained for use in preliminary testing of data collection and metric computation procedures.

Software component - general term used to refer to a software system or an element such as module, unit, data or document.

Software quality metric - a function whose inputs are software data and whose output is a single (numerical) value that can be interpreted as the degree to which software possesses a given attribute that affects its quality.

Validated metric - a metric whose values have been statistically associated with corresponding quality factor values.

B. Software Metrics Procedure

B.1. Purpose of software quality metrics

Software quality is the ratio to which software has a desired combination of qualities. This desired combination of qualities must be clearly determined; otherwise, estimation of quality is left to a hunch. For the purposes of this standard, determining software quality for a system is equal to determining a list of software quality traits necessary for that system. An apt collection of software metrics must be recognised in order to measure the software quality attributes.

The purpose of software metrics is to make assessments throughout the software life cycle as to whether the software quality requirements are being met. The use of software metrics reduces subjectivity in the assessment of software quality by providing a quantitative basis for making decisions about software quality.

However, the use of software metrics does not eliminate the need for human judgement in software evaluations. The use of software metrics within an organisation or project is expected to have a beneficial effect by making software quality more visible.

More specifically, the use of metrics within the methodology of this standard allows an organisation to:

- Achieve quality goals.

- Establish quality requirements for a system at its outset.

- Establish acceptance criteria and standards.

- Evaluate the level of quality achieved against the established requirements.

- Detect anomalies or point to potential problems in the system.

- Predict the level of quality which will be achieved in the future.

- Monitor for changes of quality when software is modified.

- Assess the ease of change to the system during product evolution.

- Normalise, scale, calibrate, or validate a metric.

To accomplish these aims, both process and product metrics should be represented in the system metrics plan.

B.2. Software quality metrics framework

The software quality metrics framework (See Figure 1 below) begins with the establishment of quality requirements by the assignment of various quality attributes. All attributes which define the quality requirements must be agreed upon by the project team, and definitions established. Quality factors, which represent management and user oriented views are then assigned to the attributes, then sub-factors, if necessary, assigned to each factor. Associated with each factor is a direct metric, which serves as a quantitative representation of a quality factor. For example, a direct metric for the factor reliability could be mean time to failure. Each factor must have one or more associated direct metrics and target values, such as one hour of execution time, that is set by project management. Otherwise, there is no way to determine whether the factor has been achieved.

At the second level of the hierarchy are the quality sub-factors, which represent software-oriented attributes that indicate quality. These are obtained by decomposing each factor into measurable software attributes. Sub-factors are independent attributes of software, and therefore may correspond to more than one factor (refer to Appendix A for further explanation). The sub-factors are concrete attributes of software that are more meaningful than factors to technical personnel, such as analysts, designers, programmers, testers, and maintainers. The decomposition of factors into sub-factors facilitates objective communication between the manager and the technical personnel regarding the quality objectives.

At the third level of the hierarchy the sub-factors are decomposed into metrics used to measure system products and processes during the development life cycle. Direct metric values (factor values) are typically unavailable or expensive to collect early in the software life cycle.

Therefore, metrics on the third level, that are validated against direct metrics, are used to estimate factor values early in the software life cycle.

The framework, in a top-down fashion, facilitates:

- Establishment of a quality requirement in terms of factors by managers early in a system's life cycle.

- Communication of the established factors to the technical personnel in terms of quality sub-factors.

- Identification of metrics that are related to the established factors and sub-factors.

On the other hand, the framework, in a bottom-up fashion, enables the managerial and technical personnel to obtain feedback by:

- Evaluating the software products and processes at the elementary metrics level.

- Analysing the metrics values to estimate and assess the quality factors.

The framework is designed to be flexible. It permits additions, deletions, and modifications of factors, sub-factors, and metrics. Each level may be expanded to several sub-levels. The framework can thus be applied to all systems and can be adapted as appropriate without changing the basic concept.

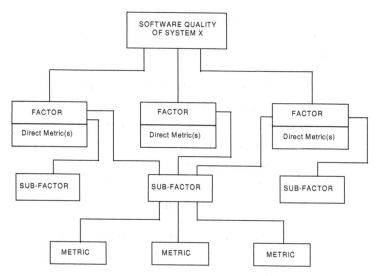

Figure 1 Software Quality Metrics Framework

B.3. The software quality metrics methodology

B.3.1. Introduction

The software quality metrics methodology is a systematic approach to establishing quality requirements and identifying, implementing, analysing and validating process and product software quality metrics for a software system. It spans the entire software life cycle and comprises five steps.

These steps shall be applied iteratively because insights gained from applying a step may show the need for further evaluation of the results of prior steps.

1. **Establish Software Quality Requirements** - A list of quality factors is selected, prioritised and quantified at the outset of system development or system change. These requirements shall be used to guide and control the development of the system and, on delivery of the system, to assess whether the system met the quality requirements specified in the contract.

2. **Identify Software Quality Metrics** - The software quality metrics framework is applied in the selection of relevant metrics.

3. **Implement the Software Quality Metrics** - Tools are procured or developed, data is collected, and metrics are applied at each phase of the software life cycle.

4. **Analyse the Software Quality Metrics Results** - The metrics results are analysed and reported to help control the development and assess the final product.

5. **Validate the Software Quality Metrics** - Predictive metrics results are compared to the direct metrics results to determine whether the predictive metrics accurately "measure" their associated factors.

The documentation produced as a result of the preceding steps is shown in Table 1

Software Metrics

Table 1. Outputs of Metrics Methodology Steps	
METRIC METHODOLOGY STEP	OUTPUT
Establish Software Quality Requirements	Quality Requirements
Identify Software Quality Metrics	Approved Quality Metrics Framework Metrics Set Cost Benefit Analysis
Implement the Software Quality Metrics	Description of Data Items Metrics/Data Item Traceability Matrix Training Plan and Schedule
Analyse the Software Quality Metrics Results	Organisation and Development Process Changes
Validate the Software Quality Metrics	Validation Results

B.3.2. Establish software quality requirements

Quality requirements shall be represented in either of the following forms:

- **Direct metric value** - a numerical target for a factor to be met in the final product. For example, mean time to failure (MTTF) is a direct metric of final system reliability.

- **Predictive metric value** - a numerical target related to a factor to be met during system development. This is an intermediate requirement that is an early indicator of final system performance. For example, design or code errors may be early predictors of final system reliability.

B.3.2.1. Identify a list of possible quality requirements

Identify quality requirements that may be applicable to the software system. Use organisational experience, and required standards, regulations, or laws to create this list. Appendix A contains sample lists of factors and sub-factors. In addition, list other system requirements that may affect the feasibility of the quality requirements. Consider acquisition concerns, such as cost or schedule constraints, warranties, and organisational self-interest. Do not rule out mutually exclusive requirements at this point. Focus on factor/direct metric combinations instead of predictive metrics.

All parties involved in the creation and use of the system shall participate in the quality requirements identification process.

B.3.2.2. Determine the actual list of quality requirements

Rate each of the listed quality requirements by importance. Importance is the function of the system characteristic and the viewpoints of the people involved. To determine the actual list of the possible quality requirements, at a minimum, follow two steps:

1. **Survey all Involved Parties** - discuss the relative priorities of the requirements with all involved parties. Have each group weigh the quality requirements against the other system requirements and constraints. Ensure that all viewpoints are considered.

2. **Create the Actual List of Quality Requirements** - resolve the results of the survey into a single list of quality requirements. This shall involve a technical feasibility analysis of the quality requirements. The proposed factors for this list may have cooperative or conflicting relationships. Conflicts between requirements shall be resolved at this point. In addition, if the choice of quality requirements is in conflict with cost, schedule or system functionality, one or the other shall be altered. Care shall be exercised in choosing the desired list to ensure that the requirements are technically feasible,

reasonable, complementary, achievable, and verifiable. All involved parties shall agree to this final list.

B.3.2.3. Quantify each factor

For each factor, assign one or more direct metrics to represent the factor, and direct metric values to serve as quantitative requirements for that factor. For example, if "high efficiency" was one of the quality requirements from the previous term, the direct metric "actual resource utilisation/allocated resource utilisation" with a value of 90% could represent that factor. This direct metric value is used to verify the achievement of the quality requirement. Without it, there is no way to tell whether or not the delivered system meets its quality requirements.

The quantified list of quality requirements and their definition shall again be approved by all involved parties.

B.3.3. Identify software quality metrics

B.3.3.1. Apply the software quality metrics framework

Create a chart of the quality requirements based on the hierarchical tree structure found in Figure 1. At this point, only the factor level shall be complete. Now decompose each factor into sub-factors. The decomposition into sub-factors shall continue for as many levels as needed until the sub-factor level is complete.

Using the software quality metric framework, decompose the sub-factors into measurable metrics. For each validated metric on the metric level, assign a target value and a critical value and range that should be achieved during development. The target values constitute additional quality requirements for the system.

The framework and the target value for the metrics shall be reviewed and approved by all involved parties.

To help ensure that metrics are used appropriately, only validated metrics (ie., either direct metrics or metrics validated with respect to direct metrics) shall be used to assess current and future product and process quality. Non-validated metrics may be included for future analysis, but shall not be included as a part of the system requirements. Furthermore, the metrics which are used shall be those which are associated with the quality requirements of the software project. However, given that the above conditions are satisfied, the selection of specific metrics as candidates for validation and the selection of specific validated metrics for application is at the discretion of the user of this standard.

Document each metric using the format shown in Figure 2 which follows.

TERM	DESCRIPTION
Name	Name of the Metric
Costs	The costs of using the metric
Benefits	The benefits of using the metric.
Impact	An indication of whether a metric may be used to alter or halt the project (ie., Can the metric be used to indicate deficient software quality.
Target Value	Numerical value of the metric that is to be achieved to meet quality requirements. Include the critical value and range of the metric.
Factors	Factors that are related to this metric.
Tools	Software or hardware tools that are used to gather and store data, compute the metric and analyse the results
Application	A description of how the metric is used and what is its area of application
Data Items	The data items (ie., input values) that

are necessary for computing the
metric values

Computation An explanation of the computation of
the metric (ie., steps involved in the
computation).

Interpretation An interpretation of the results of the
metrics computation.

Considerations Metric assumptions, appropriateness
(eg., Can data be collected for this
metric? Is the metric appropriate for
this application?)

Training Training required to implement or use
Required the metric

Example An example of applying the metric
(See Appendix C)

Validation The names of projects that have used
History the metric and the validity criteria the
metric has satisfied

References References for further details on
understanding or implementing the
metric. List of projects, project details,
etc.

Figure 2 METRICS SET

Software Metrics

A description of a Data Items is shown below.

TERM	DESCRIPTION
Name	Name of the data item
Metrics	The metrics that are associated with the data item
Definition	Unambiguous description of the data item
Source	Location where data originates
Collector	Entity responsible for collecting the data
Timing	Time(s) in life cycle at which data is to be collected (Some data items are collected more than once.)
Procedures	Methodology used to collect data (eg., automated or manual)
Storage	Location where data is stored
Presentation	The manner in which data is represented; its precision and format (eg., Boolean, dimensionless, etc.)

Sample	The method used to select the data to be collected and the percentage of the available data that is to be collected
Verification	The manner in which the collected data is to be checked for errors
Alternatives	Methods that may be used to collect the data other than the preferred method
Integrity	Who is authorised to alter this data item and under what conditions

Figure 3 Description of a Data Item

B.3.3.2. Perform a cost-benefit analysis

B.3.3.2.1. Identify the costs of implementing the metrics

Identify and document all the costs associated with the metrics in the metrics set. For each metric, estimate and document the following impacts and costs:

- **Metrics Utilisation Costs** - associated with each metric are the costs of collecting data, automating the metric value calculation (when possible), and analysing, interpreting and reporting the results.

- **Software Development Process Change Costs** - the set of metrics may imply a change in the development process.

- **Organisational Structure Change Costs** - the set of metrics may imply a change in the organisational structure used to produce the software system.

- **Special Equipment** - hardware or software tools may have to be located, purchased, adapted or developed to implement the metrics.

- **Training** - the quality assurance/control organisation or the entire development team may need training in the use of the metrics and data collection procedures. If the introduction of metrics has caused changes in the development process, the development team may need to be educated about the changes.

B.3.3.2.2.Identify the benefits of applying the metrics

Identify and document the benefits that are associated with each metric in the metrics set.

Some benefits to be considered are:

- Identify quality goals and increase awareness of the goals in the software organisation.

- Provide timely feedback useful in developing higher quality software.

- Increase customer satisfaction by quantifying the quality of the software before it is delivered to the customer.

- Provide a quantitative basis for making decisions about software quality.

- Reduce software life cycle costs by improving process efficiency based on metric data.

B.3.3.2.3.Adjust the metrics set

Weigh the benefits, tangible and intangible, against the costs of each metric. If the costs exceed the benefits of a given metric, alter or delete it from the metrics set. On the other hand, for metrics that remain, make plans for any necessary changes to the software development process, organisational structure, tools and training. In most cases it will not be feasible to quantify benefits. In these cases judgement shall be exercised in weighing qualitative benefits against quantitative costs.

B.3.3.3. Gain commitment to the metrics set

All involved parties shall review the revised metrics set to which the cost/benefit analysis has been added. the metrics set shall be formally adopted and supported by this group.

B.3.4. Implement the software quality metrics

B.3.4.1. Define the data collection procedures

For each metric in the metrics set, determine the data that will be collected and assumptions made about the data (eg., random sample, subjective or objective measure). The flow of data shall be shown from point of collection to evaluation of metrics. Describe or reference when and how tools are to be used and data storage procedures. Also identify tools form which final selection will be made. Select tools for use with the prototyping process. A traceability matrix shall be established between metrics and data items.

Identify the organisational entities that will directly participate in data collection including those responsible for monitoring data collection. Describe the training and experience required for data collection and training process for personnel involved.

B.3.4.2. Prototype the measurement process

Test the data collection and metric computation procedures on selected software. If possible, samples selected should be similar to the project(s) on which the metrics will later be used. An analysis shall be made to determine if the data is collected uniformly and if instructions have always been interpreted in the same manner. In particular, data requiring subjective judgements shall be checked to determine if the descriptions and instructions are clear enough to ensure uniform results.

In addition, the cost of the measurement process for the prototype shall be examined to verify or improve the cost analysis.

Use the results to improve the metric and descriptions of data items.

B.3.4.3. Collect the data and compute the metrics values

Following instructions in the metrics set and descriptions of data items, collect and store data at the appropriate time in the life cycle. The data shall be checked for accuracy and proper unit of measure.

Data collection shall be monitored. If a sample of data is used, requirements such as randomness, minimum size of sample and homogeneous sampling shall be verified. If more than one person is collecting the data, it shall be checked for uniformity.

Compute the metrics value from the collected data.

B.3.5. Analyse the software metrics results

B.3.5.1. Interpret the results

The results shall be interpreted and recorded against the broad context of the project as well as for a particular product or process of the project. The differences between the collected metric data and the target values for the metrics shall be analysed against the quality requirements. Substantive differences shall be investigated.

B.3.5.2. Identify software quality

Quality metric values for software components shall be determined and reviewed. Quality metric values which are outside the anticipated tolerance intervals (low or high quality) shall be identified for further study. Unacceptable quality may be manifested as excessive complexity, inadequate documentation, lack of traceability, or other undesirable attributes. The existence of such conditions is an indication that the software may not satisfy quality requirements when it becomes operational.

Since many of the direct metrics which are usually of interest (eg., reliability metrics) cannot be measured during software development, validated metrics shall be used when direct metrics are not available. Direct or validated metrics shall be measured for software components and process steps. The measurements shall be compared with critical values of the metrics. Software components whose measurements deviate from the critical values shall be analysed in detail.

The fact that a measurement deviates from a critical value does not necessarily mean that the software component will exhibit unacceptable quality during operation. This may be the case because metrics are not infallible; they are only indicators of quality. What metrics indicate about quality during development may not be quality achieved in operation.

Depending on the results of the analysis, software components shall be redesigned (acceptable quality can be achieved by redesign), scrapped (quality is so poor that redesign is not feasible), or not changed deviations for critical metric values are judged to be insignificant).

Unexpected high quality metric values shall cause a review of the software development process as the expected tolerance levels may need to be modified or the process for identifying quality metric values may need to be improved.

B.3.5.3. Make software quality predictions

During development validated metrics shall be used to make predictions of direct metric values. Predicted values of direct metrics shall be compared with target values to determine whether to flag software components for detailed analysis. Predictions shall be made for software components and process steps. Software components and process steps whose predicted direct metric values deviate from the target values shall be analysis in detail.

Potentially, prediction is very valuable because it estimates the metric of ultimate interest -- the direct metric. However, prediction is difficult because it involves using validated metrics from an early phase of the life cycle (eg., development) to make prediction about a different but related metric (direct metric) in a much later phase (eg., operations).

B.3.5.4. *Ensure compliance with requirements*

Direct metrics shall be used to ensure compliance of software products with quality requirements during system and acceptance testing. Direct metrics shall be measured for software components and process steps. These values shall be compared with target values of the direct metrics which represent quality requirements. Software components and process steps whose measurements deviate from the target values are noncompliant.

B.3.6. Validate the software quality metrics

B.3.6.1. *Purpose of metrics validation*

The purpose of metrics validation is to identify both product and process metrics that can predict specified quality factor values, which are quantitative representations of quality requirements. If metrics are to be useful, they shall indicate accurately whether quality requirements have been achieved or are likely to be achieved in the future. When it is possible to measure factor values at the desired point in the life cycle, these direct metrics are used to evaluate software quality. At some points in the life cycle, certain quality factor (eg., reliability) values are not available; they are obtained after delivery or late in the project. In these cases, other metrics are used early in a project to predict quality factor values.

The history of the application of metrics indicates that predictive metrics were seldom validated (ie., it was not demonstrated through statistical analysis that the metrics measured software characteristics that they purported to measure.) However, it is important that predictive metrics be validated before they are used to evaluate software quality. Otherwise, metrics might be misapplied (ie., metrics might be used that have little or no relationship to the desired quality characteristics).

Although quality sub-factors are useful when identifying and establishing factors and metrics, they need not be used in metrics validation, because the focus in validation is on determining whether a statistically significant relationship exists between predictive values and factor values.

Quality factors may be affected by multiple variables. A single metric, therefore, may not be sufficiently represent any one factor if it ignores these other variables.

B.3.6.2. Validity criteria

To be considered valid, a predictive metric shall demonstrate a high degree of associated with the quality factors it represents. This is equivalent to accurately portraying the quality condition(s) of a product or process. A metric may be valid with respect to certain validity criteria and invalid with respect to other criteria.

Someone in the organisation who understands the consequence of the values selected shall designate threshold values for the following:

- V - square of the linear correlation coefficient

- B - rank correlation coefficient

- A - prediction error

- P - success rate

A short numerical example follows the definition of each validity criterion. Detailed examples of the application of metrics validation are contained in the Appendix C.

B.3.6.2.1.Correlation

The variation in the quality factor values explained by the variation in the metric values, which is given by the square of the linear correlation coefficient (R) between the metric and the corresponding factor, shall exceed V, where $R^2 > V$.

This criterion assesses whether there is a sufficiently strong linear association between a factor and a metric to warrant using the metric as a substitute for the factor, when it is infeasible to use the latter.

For example, the correlation coefficient between a complexity metric and the factor reliability may be .8. The square of this is .64. Only 64% of the variation in the factor is explained by the variation in the metric. If V has been established as .7, the conclusion would be drawn that the metric is invalid (ie., there is insufficient association, or correlation between the metric and reliability). If this relationship is demonstrated over a representative sample of software components, the conclusion could be drawn that the metric is invalid.

B.3.6.2.2.Tracking

If a metric M is directly related to a quality factor F, for a given product or process, then a change in a quality factor value from F_{T1} to F_{T2}, at times T1 and T2, shall be accompanied by a change in metric value from M_{T1} to M_{T2}, which is the same direction (eg., if F increases, M increases). If M is inversely related to F, then a change in F shall be accompanied by a change in M in the opposite direction (eg., if F increases, M decreases). To perform this test, compute the coefficient of rank correlation (r) from n paired values of the factor and the metric. Each of the factor/metric pairs is measured at the same point in time, and the n pairs of values are measured at n points in time. The absolute value of (r) shall exceed B.

This criterion assesses whether a metric is capable of tracking changes in product or process quality over the life cycle.

For example, if a complexity metric is claimed to be a measure of reliability, then it is reasonable to expect a change in the reliability of a software component to be accompanied by an appropriate change in metric value (eg., if the product increases in reliability, the metric value should also change in a direction that indicates the product has improved). That is, if MTTF is used to measure reliability and is equal to 1000 hours during testing (T1) and 1500 hours during operation (T2), a complexity metric whose value is 8 in T1 and 6 in T2, where 6 is 'better' than 8 (ie., complexity has decreased), is said to track reliability for this software component. If this relationship is demonstrated over a representative sample of software components, the conclusion could be drawn that the metric can track reliability (ie., indicate changes in product reliability) over the software life cycle.

B.3.6.2.3.Consistency

If factor values F1, F2, ..., Fn, corresponding to products or processes 1, 2, ..., n, have the relationship F1 > F2 > ..., Fn, the corresponding metric values shall have the relationship M1 > M2 > ..., Mn. To perform this test, compute the coefficient of rank correlation (r) between paired values (from the same software components), of the factor and the metric; |r| shall exceed B.

This criterion assesses whether there is consistency between the ranks of the factor values of a set of software components and the ranks of the metric values for the same set of software components. Thus this criterion is used to determine whether a metric can accurately rank, by quality, a set of products or processes.

For example, if the reliability of software components X, Y and Z, as measured by MTTF, is 1000, 1500 and 800 hours, respectively, and the corresponding complexity metric values are 5, 3 and 7, where low metric values are 'better' than high values, the ranks for reliability and metric values, with '1' representing the 'highest' rank, are as follows:

Software Component	Reliability Rank	Complexity Metric Rank

Y	1	1
X	2	2
Z	3	3

If this relationship is demonstrated over a representative sample of software components, the conclusion could be drawn that the metric is consistent and can be used to rank the quality of software components. For example, the ranks could be used to establish priority of testing and allocation of budget and effort to testing (ie., the 'worst' software component would receive the most attention, largest budget and most staff).

B.3.6.2.4.Predicability

If a metric is used at time T1 to predict a quality factor for a given product or process, it shall predict a related quality factor F_{pT2} with an accuracy of:

$$\left| \frac{F_{aT2} - F_{pT2}}{F_{aT2}} \right| < A$$

where F_{aT2} is the actual value of F at time T2.

This criterion assesses whether a metric is capable of predicting a factor value with the required accuracy.

For example, if a complexity metric is used during development to predict the reliability of a software component during operation (T2) to be 1200 hours MTTF (F_{pT2}) and the actual MTTF that is measured during operation is 1000 hours (F_{aT2}), then the error in prediction is 200 hours, or 20%. If the acceptable prediction error (A) is 25%, prediction accuracy is acceptable. If the ability to predict is demonstrated over a representative sample of software components, the conclusion could be drawn that the metric can be used as a predictor of reliability. For example, prediction could be used during development to identify those software components that need to be improved.

B.3.6.2.5.Discriminative Power

A metric shall be able to discriminate between high quality software components (eg., high MTTF) and low quality software components (eg., low MTTF). For example, the set of metric values associated with the former should be significantly higher (or lower) than those associated with the latter.

This criterion assesses whether a metric is capable of separating a set of high quality software components from a set of low quality components. This capability identifies critical values for metrics which can be used to identify software components which may have unacceptable quality. The Mann-Whitney Test and the Chi-Square Test for

Differences in Probabilities (Contingency Tables) can be used for this validation test.

For example, if all software components with a complexity metric value of > 10 (critical value) have a MTTF of 1000 hours and all components with a complexity metric value equal to or less than 10 have a MTTF or 2000 hours, and this difference is sufficient to pass the statistical tests, then the metric separates low from high quality software components. If the ability to discriminate is demonstrated over a representative sample of software components, the conclusion could be drawn that the metric can discriminate between low and high reliability components for quality assurance and other quality functions.

B.3.6.2.6.Reliability

A metric shall demonstrate the above correlation, tracking, consistency, predicability and discriminative power properties for P percent of the application of the metric.

This criterion is used to ensure that a metric has passed a validity test over a sufficient number or percentage of applications so that there will be confidence that the metric can perform its intended function consistently.

For example, if the required 'success rate' (P) for validating a complexity metric against the Predicability criterion has been established as 80%, and there are 100 software components, the metric shall predict the factor with the required accuracy for at least 80 of the components.

B.3.6.3. Validation procedure

Metrics validation shall include the following steps.

B.3.6.3.1.Identify the quality factors ample

These factor values (eg., measurements of reliability), which represent the quality requirements of a project, were previously identified and collected and stored. For validation purposes, draw a sample from the metrics database.

B.3.6.3.2.Identify the metrics sample

These metrics (eg., design structure) are used to predict or to represent quality factor values, when the factor values cannot be measured. The metrics were previously and their data collected and stored, and values computed from the collected data. For validation purposes, draw a sample from the same domain (eg., same software components) of the metrics data base.

B.3.6.3.3.Perform a statistical analysis

The tests described under Validity Criteria shall be performed.

Before a metric is used to evaluate the quality of a product or process, it shall be validated against the criteria. If a metric does not pass all of the validity tests, it shall only be used according to the criteria prescribed by those tests (eg., if it only passes the 'Tracking' validity test, it shall be used only for 'tracking' the quality of a product or process).

B.3.6.3.4.Document the results

This shall include the direct metric, predictive metric, validation criteria and numerical results, as a minimum.

B.3.6.4. Additional requirements

B.3.6.4.1.The need for re-validation

It is important to re-validate a predictive metric before it is used for another environment or application. As the software engineering process changes, the validity of metrics changes. Cumulative metric validation values may be misleading because a metric that has been valid for several uses may become invalid. It is wise to compare the one-time validation of a metric with its validation history to avoid being misled. The following statements of caution should be noted:

- A validated metric may not necessarily be valid in other environments or future applications.

- A metric that has been invalidated may be valid in other environments or future applications.

B.3.6.4.2.Confidence in analysis results

Metrics validation is a continuous process. Confidence in metrics is increased over time and with a variety of projects as metrics are validated, the metrics data base increases and sample size increases. Confidence is not a static, one-time property. If a metric is valid, confidence will increase with increased use (ie., the correlation coefficient will be significant at decreasing values of the significance level). Greatest confidence occurs when the metrics have been tentatively validated based on data collected from previous projects. Even when this is the case, the validation analysis will continue into future projects as the metrics data base and sample size grow.

B.3.6.4.3.Stability of environment

To the extent practicable, metrics validation shall be undertaken in a stable development environment (ie., where the design language, implementation language or program development tools do not change over the life of the project in which validation is performed). In addition, there shall be at least one project in which metrics data have been collected and validated prior to application of the predictive metrics. This project shall be similar to the one in which the metrics are applied with respect to software engineering skills, application, size and software engineering environment.

Validation and application of metrics shall be performed during the same life cycle phases on different projects. Example: if metric X is collected during the development phase of project A and the sae values are later validated with respect to direct metric Y, whose values are collected during the operations phase of project A, the metric X shall be used during the development phase of project B to assess direct metric Y with respect to the operations phase of project B.